ON TIDES

AND

TIDAL ACTION IN HARBORS.

BY

PROFESSOR J. E. HILGARD

OF THE

UNITED STATES COAST SURVEY.

REPRINTED FROM THE SMITHSONIAN REPORT FOR 1874.

WASHINGTON:
GOVERNMENT PRINTING OFFICE.
1875.

ON TIDES AND TIDAL ACTION IN HARBORS.*

By Professor J. E. Hilgard, of the U. S. Coast Survey, Washington, D. C.

Ladies and Gentlemen: I propose to engage your attention this evening with the subject of the tides of the ocean and the influence exerted by tidal currents on our harbors. I shall first briefly describe the phenomena of the tides as they present themselves to an observer, then consider the physical causes to which these phenomena are due, next examine more in detail the phases of the tide on our own coasts, and finally describe the tidal hydraulics of the magnificent harbor of New York.

The most obvious change in the surface of the ocean to be noticed upon our shores is the alternate rising and falling regularly twice in every day. Closer attention will show that the tides of each day occur somewhat later than those of the preceding day, the average time of retardation being fifty-two minutes, and that this retardation corresponds to that of the moon. It will pass as a fair approximation to say, that it is high water at New York with a southeast moon, or similarly for New Castle, on the Delaware, that high water occurs when the moon is south In fact, so closely is the time of tide connected with the position of the moon, that in order to give the time of high water upon any day approximately it is customary to state the time of high water on the days of the new and full moon, when the moon passes the meridian at twelve o'clock, nearly. This time is called the "establishment of the port." Then, to find the time of high water on any other day, it is only necessary to add the "establishment" to the time of the moon's meridian passage on that day. On closer examination, it will be found that the interval between the time of the moon's passage over the meridian and the time of high water, called the *luni-tidal interval*, varies with the moon's age very sensibly. Moreover, the elevation at high water and depression at low water will not always be the same, but will be greatest about the times of new moon and full moon, and least about the first and third quarters. The details of these variations will be best traced out in connection with the explanation of their causes, to which we will now proceed.

The popular explanation of the tides, as depending on the law of gravitation, is sufficiently simple, although the complete mathematical investigation of the subject, by which we should be enabled to predict their

* Delivered before the American Institute, January 27, 1871, with revision.

occurrence and magnitude for any place, is encompassed with difficulties, from causes to which we shall hereafter revert.

If we conceive the earth to be wholly, or for the greater part, covered with water and subject to the attraction of the sun, the force of which varies inversely as the square of the distance, it will be obvious, that while the whole earth will fall toward the sun with a velocity proportioned to the aggregate attraction upon its solid portions, (which is the same as if all the matter were collected at its center,) the water nearest to the sun being accelerated by a greater force, and being fluid, will approach the sun more rapidly than the solid core. It will thus run from all sides into a protuberance beyond the form of equilibrium of the earth's attraction and rotation, until the pressure of the elevated mass equals the difference in the attraction of the sun. Moreover, a similar protuberance will be formed on the side opposite to the sun, since the particles of water, being solicited by a less force than the solid core, will fall more slowly toward the sun, and as it were remain behind. Nor does the fact that on the average the earth does not lessen its distance from the sun, in the least invalidate the force of this reasoning; for the deviations from the tangential motion of the earth in its orbit are precisely those which the earth would move through if falling toward the sun unaffected by any other impulse.

The same considerations hold good in regard to the attraction of the moon upon the earth and the waters surrounding it; for although we are in the habit of considering the moon as simply revolving about the earth, it must be remembered that the attraction is mutual, that both bodies describe orbits about their common center of gravity, and that while the moon obeys the attractive force of the earth, the latter equally follows that of the former, by which it is at every instant of time drawn from the path which it would pursue if that influence did not exist by an amount precisely equal to the fall corresponding to the moon's attractive force.

As a necessary consequence of the elevation of the water in the regions nearest to and most remote from the attracting body, there must be a corresponding depression below the mean level of the sea at points distant ninety degrees from the vertices of the protuberances, or at the sides of the earth, as seen from the sun or moon. If the latter bodies maintained a constant position with respect to the earth, the effect would therefore be to produce a distortion of figure in the ocean-surface, (assumed to cover the whole earth,) having the form of a slightly elongated ellipsoid, the two vertices of which would be the one precisely under, the other precisely opposite to, the points at which the disturbing body is vertical. This, however, is not the case; for by the rotation of the earth, and the motion of earth and moon in their orbits, the direction of the disturbing forces is constantly changing with respect to any point on the earth's surface. New points arrive at every instant under the zenith and nadir of either luminary, and thus it is that waves are produced which follow them round the globe. The highest points of these waves will remain far behind the verticals of the disturbing bodies, because the inertia and friction of the water prevent the rapid change of form required, and because, although the elevating force is greatest under the vertical, it still continues to act in the same direction for some hours after the passage of the luminary, with but little diminished force.

This retardation, which would be sensible under the simple supposition of an uninterrupted ocean covering the earth's surface, becomes very considerable under the actual circumstances of the case. The depth of the sea varies so much, and the form of its basin, taken as a whole, is so interrupted by the land, that no regular progressive movement of the tide-wave can take place, except in the great Southern Ocean. At all points on the coast the phases of the tide will follow the periodicity of the forces causing them, but at each point, at a greater or less interval from the culmination of the sun or moon, according to its local position, and the more or less circuitous course taken by the tide-wave to reach it. This interval and the actual rise and fall of the tide must be determined for each place by special observation.

LUNI-SOLAR PHASES OF THE TIDES.

The close relations which the times of high water bear to the times of the moon's passage show that the moon's influence in raising the tides must be much greater than the sun's. In fact, while the *whole* attraction of the sun upon the earth far exceeds that of the moon, yet owing to the greater proximity of the latter, the *difference* between its attraction at the center of the earth and at the nearest or most remote point of its surface, which difference alone produces the tides, is about two and a half times as great as the difference of the sun's attraction at the same points.

SEMI-MONTHLY INEQUALITY.

We will now consider the particular phases resulting from the combination of the lunar and solar tides, and from the varying positions of those bodies. There will be two complete lunar tides in every lunar day of twenty-four hours fifty-two minutes, and also two complete solar tides in every mean solar day of twenty-four hours. These are known as the semi-diurnal tides, and constitute the principal variations of the sea-level. The combined effect of these two fluctuations will be most readily understood by reference to the annexed diagram, in which the lunar tide is represented by dashes, the solar by dots, and the combined or actual tide by a full line. At the time of syzygies, or full and change of the moon, the effects of both sun and moon combine together to produce the *spring-tides*, when high water is higher and low water is lower than at mean tides by the amount of the solar tide. At quadratures the high water of the sun will combine with the low water of the moon to produce a less fall, and the low water of the sun with the high water of the moon to produce a less rise than at mean tides; and we have the *neap-tides*, the range of which is less than the mean range by the amount of the solar tide. Thus, at New York, the rise and fall at syzygies is 5.4 feet, at quadrature 3.4 feet, the former being the sum,

SEMI-MONTHLY INEQUALITY

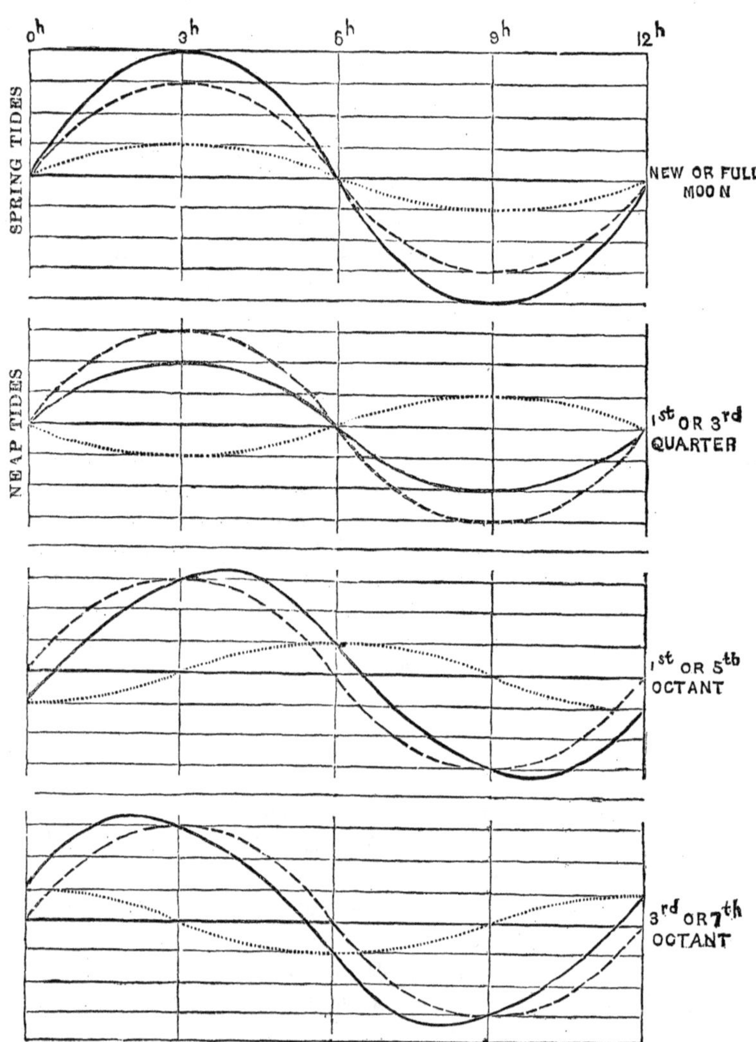

the latter the difference of the lunar and solar tides, whence we obtain for the effect of the moon 4.4 feet and for that of the sun one foot, or a ratio of forty-four to ten. This proportion does not prove to be the same in all parts of the world, and even varies considerably in places not far distant from each other. At Boston the heights are 11.3 and 8.5 feet, respectively, giving a proportion of seven to one. On the Atlantic coast of the United States it averages about five to one, while on the east side of the Atlantic Ocean, on the coasts of France and England, it is in many parts as three to one. These differences are to be ascribed to the fact that the shore and harbor tides which we observe have in every instance acquired a greater magnitude than the ocean tides, in consequence of the wave having passed over a sloping bottom and having been greatly retarded by the effect of friction. A comparison of the range of spring and neap-tides, therefore, will not serve as a correct measure of the relative effect of the sun and moon, unless the effect of friction were taken into consideration, which we are at present unable to do for want of a complete knowledge of the configuration of the bottom.

The interval between the moon's meridian passage and the time of high water is subject to a variation similar to that of the height. On the day after the spring-tides, the top of the solar tide-wave will be nearly an hour in advance of the lunar tide-wave, and the two waves will combine to make high water earlier than the moon's alone would bring it. It will continue to be earlier until the moon's transit is later by three hours, or in the first octant. It then falls back until it is latest in the third octant, and again advances, until, at the next spring-tides, it reaches its mean period. The mean of all the luni-tidal intervals for half a month at a port its called its *mean establishment*, which is used for finding the time of high water on any given day; and tables are constructed from observations at the principal ports for finding the correction for semi-monthly inequality due to the moon's age. Thus, for New York, the mean luni-tidal interval is 8h. 13m., and its least and greatest values are 7h. 52m. and 8h. 35m. On the Atlantic coast of the United States the range of this inequality is about three-quarters of an hour; on the coasts of France and Great Britain it often exceeds one and a half hours.

DIURNAL INEQUALITY.

The next variation of the tides to be considered is that dependent on the moon's declination. Were that body constantly in the plane of the equator, the highest points of the tide-waves would also be in that plane, and would consequently produce a series of equal tides at any place either north or south of the equator. But it is evident that, when the moon ascends to the north, the vertex of the tide-wave will tend to follow it, giving the highest point of one tide in the northern, and the highest point of the opposite tide in the southern, hemisphere. Consequently, when the moon has a northern declination, the tide at any place in the northern hemisphere caused by its upper transit will be higher than that caused by its lower transit. (See diagram of diurnal inequality.) This variation in the heights has a period of one lunar day, and is called the *diurnal inequality;* it reaches its maximum when the moon is at its greatest northern or southern declination, and disappears when it is on the equator, and consequently has a half-monthly period. The variations of height from this cause produce a corresponding inequality in the times of high water. The sun's declination affects the tides in a similar manner, but the amount of the disturbance is very small, and its period extends over half a year. In long series of observations its effect is nevertheless well marked, both in height and time. The diurnal inequality, depending upon the moon's declination, is, on the other hand, quite sensible, and in many places constitutes a prominent feature of the tides, as on the Pacific coast of North America.

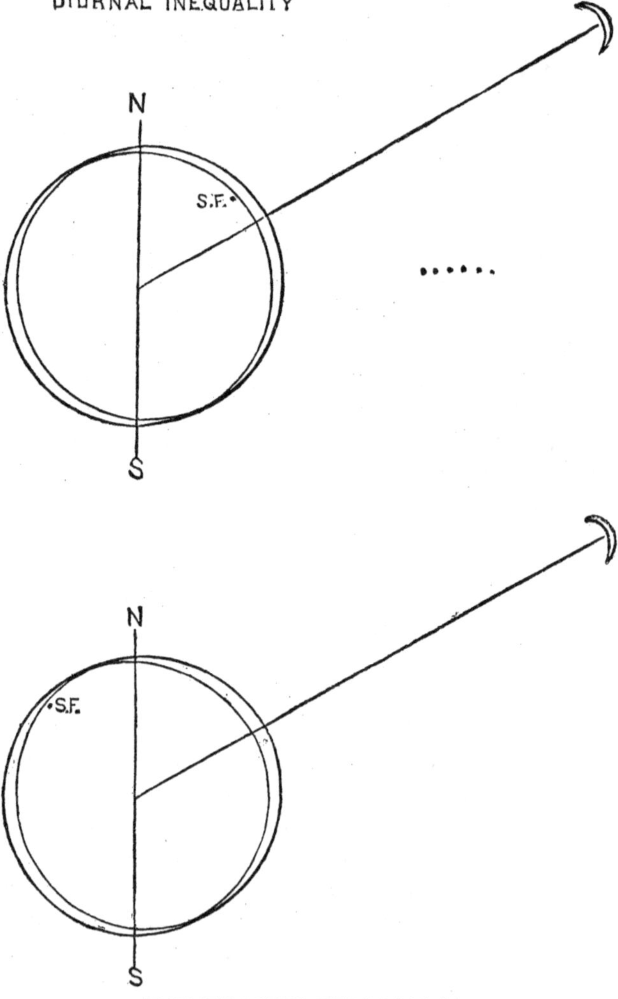

PARALLACTIC INEQUALITY.

A further cause in the variation of the height of the tides is the variation of the distances of the sun and the moon by reason of the ellipticity of their orbits. The efficacy of a heavenly body in raising tides is shown by theory to be inversely proportional to the cube of the distance. Hence the efficacy of the sun will fluctuate between the extremes nineteen and twenty-one, taking twenty for its mean value, and that of the moon between forty-three and fifty-nine. Taking into account this cause of difference, the highest spring-tide will be to the lowest neap as 59+21 to 43—19, or as eighty to twenty-four, or ten to three; leaving out of consideration the local circumstances of access and depth, which, as we have stated, modify those proportions in a marked degree.

TYPE CURVES.

The three principal forms of tides are illustrated in the annexed diagram, which exhibits the tides at New York, San Francisco, and Galveston for two days from actual observation. Of these, that for San Francisco may be taken as the normal type, showing the diurnal inequality, while that at New York, as at other ports on the Atlantic coast, is not sensibly affected by it. The explanation of this feature is probably to be found in the supposition that the tide-wave which advances up into the Atlantic Ocean from the continuous tide in the Southern Ocean arrives on our shores twelve hours later than the direct tide-wave which crosses the Atlantic from east to west. In this way the diurnal inequality will be eliminated by the superposition of the two tides, the greater high water of the former coinciding with the lesser of the latter, and *vice versa*, leaving the semi-diurnal tides of equal height.

TIDE REGISTERS

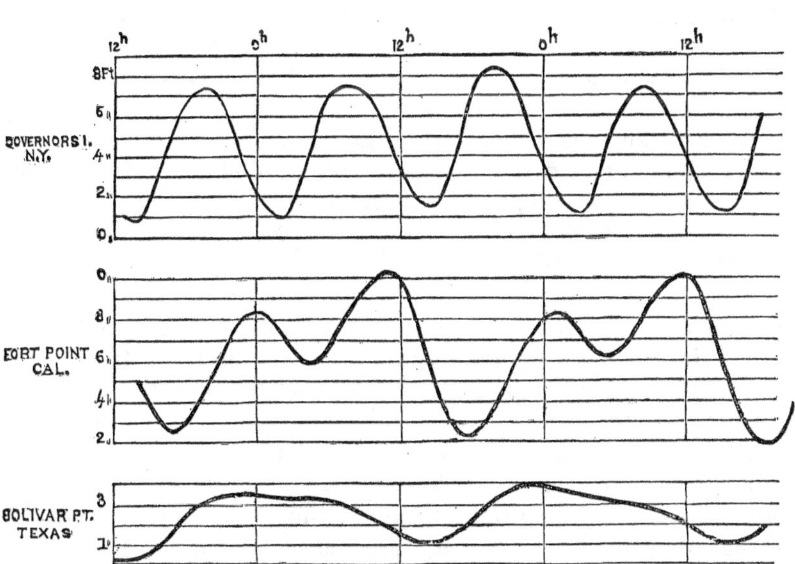

The tide at Galveston, on the other hand, furnishes a case of the elimination of the semi-diurnal tide, leaving as a residual only the diurnal inequality. It is to be presumed in this instance that the tides reaching Galveston through the straits of Florida and through the passage between Cuba and Yucatan differ by six hours in their periods, causing the low water of one to coincide with the high water of the other, thus sensibly destroying the semi-diurnal tides, except in so far as they are unequal. This leaves a small tide outstanding, having substantially the form of the diurnal inequality, and producing the appearance of the "single-day tide," or one high and one low water in every twenty-four hours. This residual fluctuation is well marked at times when the moon's declination is considerable on either side of the equator, but disappears almost entirely when the moon is near the equator, since, at such times, the diurnal inequality disappears. Tides of this class have always a small range; in the Gulf of Mexico they rarely exceed two and a half feet, and the average rise and fall is but one and a half foot.

The tides on the coasts of the United States have been specially investigated by Professor Bache, the late Superintendent of the American Coast Survey. In connection with that work he organized an extensive system of exact tidal observations, for the purpose of ascertaining the complicated laws which govern the tides of the seas that wash our shores. It will be readily understood that in order to separate the effects of the different causes which modify the phenomena, it is not sufficient to observe merely the heights and times of high and low water, but that a continuous record of the tides is necessary, as the inequalities are constantly shifting their place and magnitude.

TIDE-GAUGES.

For this purpose a self-registering tide-gauge is used, by which a continuous curve, representing the successive changes in the height of water, is traced on paper, moved by clock-work, by a pencil actuated by the rising and falling of a float in a vertical box, to which the tide has free access. The time-scale is such that every hour is represented by one inch, and is pricked into the paper by points on the cylinder which moves the paper forward. The scale of heights is so adapted to the range of the tide at the place of observation that the extreme range of the curve will not exceed the width of the sheet—twelve inches. A continuous sheet, sufficient for the record of a whole month, is put on the tide-gauge at one time. A complete description of this instrument will be found in the United States Coast Survey Report for 1853. [The lecturer illustrated the construction of several tide-gauges by means of diagrams.]

In northern ports interruptions are experienced in winter from the float-box becoming clogged with ice, and various devices have been resorted to for overcoming this difficulty. One of the most effective has been that of maintaining a temperature above freezing within the float-box by means of a simple heating-apparatus. An arrangement of this kind has actually been used on the Fox Islands, in Penobscot Bay. A stream of water flows slowly from an elevated hogshead through a coil in a large stove, passes down to the bottom of the float-box and up again into another hogshead, from which it is pumped up every day by the observer into the first one. As but a small elevation of temperature is necessary, this arrangement has proved quite sufficient.

Another arrangement, devised by Mr. Batchelder, of Boston, and called by him an "Arctic tide-gauge," is in use at Boston, and has compared well with the ordinary float-gauge. It consists of a strong iron tube, about four inches in diameter, firmly bolted to a wharf or pile. It is open at the top, and has at the lower end a nipple, to which an India-rubber bag is fastened; the length of the tube being sufficient to allow the elastic bag to be always submerged at the lowest stage of the tide. The bag is supported by a suitable shelf or cage, and is filled with glycerine, which is poured in at the top of the tube. When in this condition the glycerine rises and falls within the iron tube in proportion to the varying height and pressure of the column of water above the rubber bag, the difference in the height of the two columns being in proportion to the difference of the specific gravity of the water and the glycerine. The parts above described insure protection against floating ice, and prevent congelation within the iron tube.

A copper tube about three inches in diameter, closed at the bottom and open at the top, is placed within the iron tube, and floats in the glycerine; if left free, it would rise and fall with the changing level of this liquid. The length of the central tube is a little greater than the whole range of the tide.

Near the upper end of the outer tube there are three spiral springs, fixed at the top and united at the bottom by a plate or disk, from which the central copper tube is suspended. From a stem fixed to the central tube or float, and moving with it, a string or chain leads over a single pulley, and gives horizontal motion to the pencil-carriage of the recording-apparatus.

The distance that the central tube is to move vertically is adjusted to agree with the required range of the pencil upon the record-paper by placing within it suitable weights.

As the glycerine rises or falls in the annular space between the iron tube and the central float, the spiral spring at the top is more or less extended, the extension being uniform on account of the cylindrical form of the float.

It is not necessary that the India-rubber bag be inclosed in a perforated box, for the purpose of preventing oscillation, as it is always submerged, and the pressure upon it is equal to the weight of the column of water, having its base at the bag, and its summit at the mean level of the surface-waves.

A tide-gauge, for observations on an open coast, has been devised by Mr. Henry Mitchell, of the Coast Survey. The graduated scale on the float is read from the shore by means of a spy-glass, the top of the tube serving as index-mark.

PREDICTION OF TIDES.

Self-registering tide-gauges have been kept in operation for a number of years at different points on both coasts of the United States, in order to obtain from them the data for predicting the tides; and as a result, tide-tables have been published by the Coast Survey for some years past, giving in advance the times and heights of high and low water for all the principal ports in the United States for every day in the year. In addition to this, the differences are given by which to find the same for intermediate ports.

A very elaborate discussion of the tides observed at Boston during nineteen years, a full lunar cycle, has been made by Mr. William Ferrel, of the Coast Survey, and has resulted in representing the actual tides with unlooked-for precision. By the introduction of the consideration of friction Mr. Ferrel has also succeeded in deriving a value for the mass of the moon, which appears to compete in exactness with the values obtained by astronomical methods. It is one seventy-seventh part of that of the earth.

EARTHQUAKE-WAVES.

The tide-gauges being in continuous operation, all other fluctuations of the ocean-level besides those produced by the tides are likewise registered. The tide-curves of the western coast are frequently found indented by fluctuations arising from earthquakes. A remarkable instance of this kind is given in the annexed diagram of earthquake-waves, which

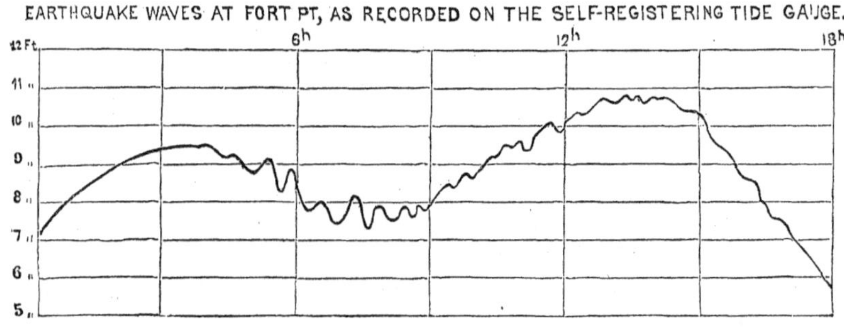

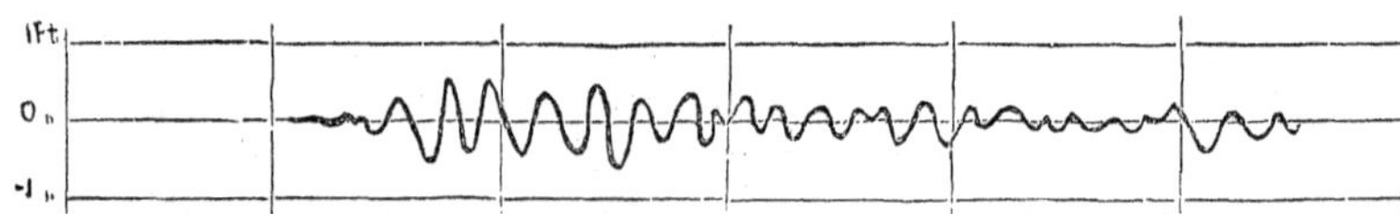

recorded the earthquake that destroyed the city of Simoda, in Japan, in December, 1854. The upper curve is a reduction from the tide-gauge register, while the lower shows the earthquake-waves separated from e tidal wave. The time required for the transmission of the sea-waves om Simoda to San Francisco was twelve hours and thirty-six minutes. he distance being 4,500 miles, the transmission of the wave was at an aver-ge rate of 360 miles per hour. The theory of wave-motion teaches us that is velocity will be attained by a free-moving wave in a depth of 1,440 thoms, which may be taken as the average depth of the Pacific Ocean etween Japan and California. It will be observed that the crests of e waves occur at intervals of about twenty-three minutes, correspond-g to a length, from crest to crest, of 150 miles. The height when e waves arrived at San Francisco was about eighteen inches from ollow to crest, the high waves caused by the original impulse having radually flattened out to that form in their transmission across the cean.

The great earthquake which occurred in Peru, in August, 1868, was kewise recorded on the tide-gauges at San Diego, San Francisco, and storia. The fluctuation of the ocean was so great in this instance as o be very sensible to casual observation, and was noted in Australia, t the Sandwich Islands, and at Kodiak, in Alaska. The data obtained rom these observations, combined with the result before mentioned, ndicate that the average depth of the Pacific Ocean is about 1,800 athoms.

MOVEMENT OF TIDAL WAVES.

The waves above described, originating with an impulse at one definite oint, and propagated freely through the ocean in every direction, with a velocity depending upon the square root of the depth of the sea, may serve as good illustrations of the manner in which tides are propagated through sounds, bays, and rivers. The following table gives the rate of motion for different depths:

Depth in feet		10	Miles per hour		12.2
"	"	60	"	"	30.0
"	"	100	"	"	38.7
"	"	1,000	"	"	122.3
"	"	6,000	"	"	299.5

That movement of the ocean, however, which we have designated by the name of tide-wave, does not partake of the nature of a wave in the common acceptation of the term, but it is rather to be conceived as a general movement of the water toward a point under the attracting body, and again away from it. Its periodicity is strictly dependent upon that of the attracting body. The velocity of the movement is about 1,000 miles per hour on the equator; it extends to the bottom of the ocean, the depth of which is inconsiderable compared with the radius of the earth. It is not attended by a sensible elevation of the water in mid-ocean; and in this respect the characteristic of what we call a wave is absent. The movement may be likened to that of an impulse given to a very long rigid bar, as of iron. In this case, a sensible time will be required for the transmission of the impulse from one end to the other, and during its transmission the particles will successively approach to each other, by which an infinitesimal elevation and subsidence, after the manner of a wave, will be produced. In the same way the trans-

TIMES AND HEIGHTS OF TIDES ON ATLANTIC COAST OF UNITED STATES

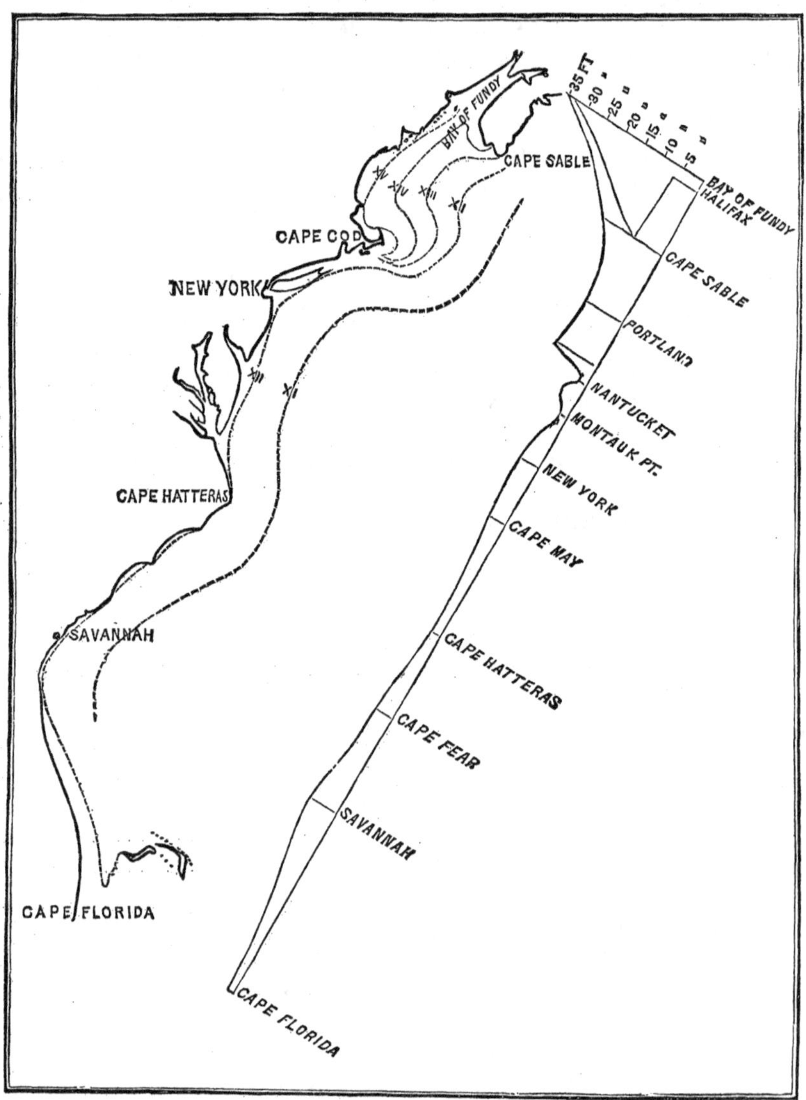

mission of the movement through the incompressible water of the sea is attended with an infinitesimal elevation and recession; but when the movement reaches shallow water, in approaching the shores, the horizontal motion is partly translated into vertical motion upon the sloping bottom; and it is thus that the tides attain sensible vertical height.

Now, where a bay or indentation of the coast presents itself, opening favorably to the tide-wave thus developed, and decreases in width from its entrance toward its head, the tide rises higher and higher from the mouth upward. This is due to the concentration of the wave by the approach of the shores and to the gradual shoaling of the bottom.

This effect is strikingly illustrated by a generalization of the heights of the tides on the Atlantic coast of the United States. That coast presents, in its general outline, as represented in the annexed diagram, three large bays: the great southern, from Cape Florida to Cape Hatteras; the great middle, from Cape Hatteras to Nantucket; and the great eastern, from Nantucket to Cape Sable, now known as the Gulf of Maine. It will be seen that the tide-wave arrives at about the same time at the headlands, Cape Florida, Cape Hatteras, Nantucket, and Cape Sable, and that at those points the height is inconsiderable compared with the rise at the head of the several bays. Thus, at Cape Florida the mean rise and fall is only one and one-half of a foot; at Hatteras, but two feet; while at the intermediate entrance to Savannah it reaches seven feet, declining in height toward both capes. Again, at the head of the middle bay, in New York Harbor, it reaches five feet, while on the southeast side of Nantucket Island it is little over one foot. The configuration of the eastern bay is less regular, and the correspondence of heights is not so obvious. The recess of Massachusetts Bay is well marked, the increase in height reaching ten feet at Boston and Plymouth. Rolling on eastward along the coast of Maine, it constantly increases; but the most striking effect of the convergence of shores is exhibited in the Bay of Fundy. At St. John's the mean height of tide is nineteen feet, and at Sackville, in Cumberland Basin, thirty-six feet, attaining to fifty feet and more at spring-tides.

When the wave leaves the open sea, its front slope and rear slope are equal in length and similar in form, but as it advances into a narrow channel, bay, or river, its front slope becomes short and steep, and its rear slope becomes long and less inclined. Hence arises the fact that at a station near the sea, the time occupied by the rise is equal to that occupied by the descent; but at a station more removed from the sea, the rise occupies a shorter time than the descent. Thus, in Delaware Bay and River we have the following relations of the duration and height of rise and fall:

Station.	Mean rise and fall.	Luni-tidal interval.	Mean duration of—	
			Flood-tide.	Ebb-tide.
	Feet.	*h. m.*	*h. m.*	*h. m.*
Delaware breakwater	3.5	8 0	6 18	6 8
Egg Island light	6.0	9 4	5 56	6 30
New Castle	5.5	11 53	5 24	7 2
Philadelphia	6.0	13 44	4 52	7 34

An examination of this table will show, besides the marked increase in the height of the tide due to the contraction of the shores from the capes up to New Castle, a subsequent loss from friction in a narrow channel of nearly uniform character, and correspondingly a rapid propagation of the tide-wave through the deep water of the bay, and a comparatively slow movement along the narrower channel of the river. At the mouth of the bay the duration of rise exceeds that of fall by ten minutes, while at Philadelphia it is less by two hours forty-two minutes. When the tide is very large compared with the depth of water, this inequality becomes very great; thus, in the Severn River, at Newnham, above Bristol, England, the whole rise of eighteen feet takes place in one and a half hours, while the fall occupies ten hours.

TIDAL CURRENTS.

The agency of tidal currents in producing changes in the entrances of bays and harbors is a subject of the first importance to commerce and navigation, and has received full attention in the prosecution of the American coast survey. The laws according to which the changes take place require to be studied by long-continued observation, and when the change is for the worse, the means of counteracting it must be pointed out.

As on the average the same amount of water moves inward and outward with the flood and ebb tides, we might readily suppose that the same amount of material is transported either way, and that no important change would take place in the configuration of the bottom. But the operation of the flood-stream is very different from that of the ebb-stream. We have, as a general feature, an interior basin of some extent, communicating with the sea by a comparatively narrow passage. The flood-stream, therefore, running with considerable velocity through this channel, will, as it enters the basin, spread out and become slow, depositing the sand and mud it is charged with, and making extensive flats or shoals opposite the entrance. The ebb-stream runs slowly over the flats from all directions toward the opening without removing much of the deposit, and gradually concentrates in definite narrow channels, which it scoops out, and the depth of which will depend in a great degree on the proportion of the area of the basin to the outlet, or, in other terms, on the difference of level which will be reached during the ebb between the basin and the ocean, which determines the greatest velocity and transporting power reached by the ebb-stream.

On the bars of most of the sand-barred harbors on our southern coast, the place and direction of the channel are frequently changed during violent storms; when the direction of the waves happens to be oblique to that of the channel, or when the sea runs directly upon the channel, the depth of water may be considerably diminished for the time being by the sand rolled up by the waves. But in all these cases it is found that the normal depth is speedily restored by the scour of the ebb-tide, which depends upon the unchanged factors of area and form of basin, height of tide, and character of the material forming the bar.

EFFECT OF SINKING STONE-FLEET ON CHARLESTON BAR

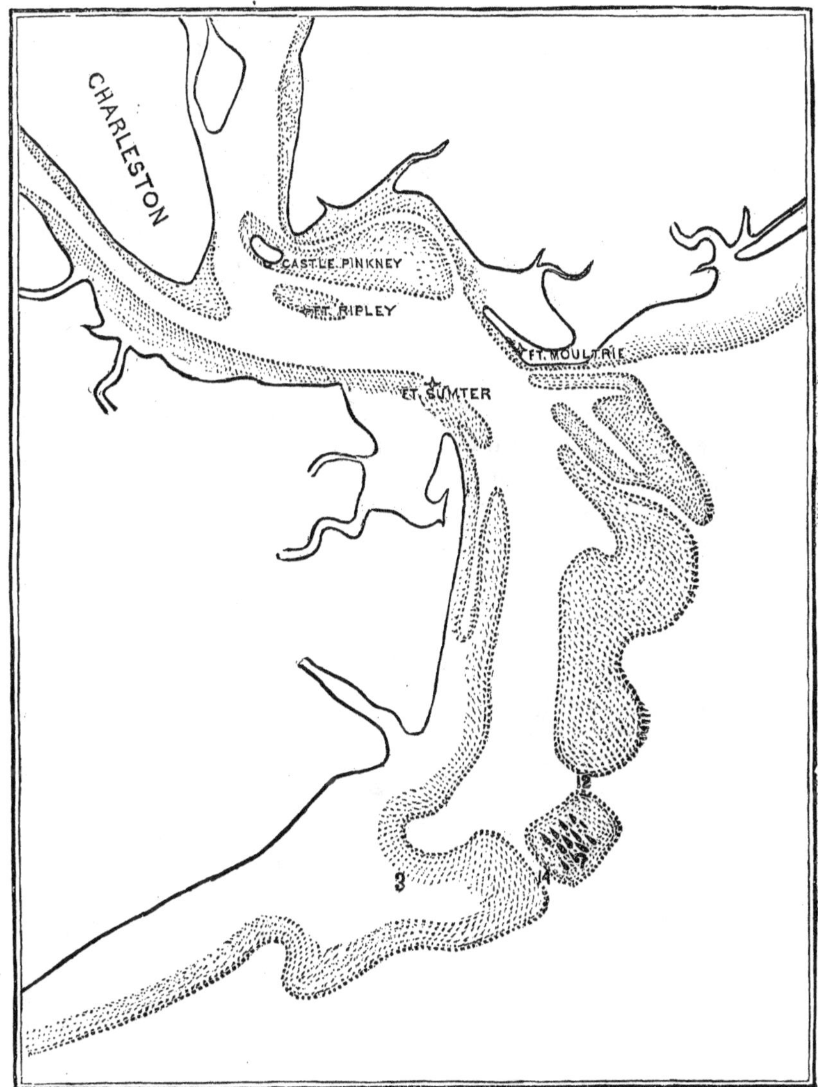

An interesting instance of this maintenance of the depth of channels from a determinate tidal basin is furnished by the effects of the obstructions placed in the channel over Charleston Bar during the war of the rebellion. On the accompanying diagram is seen the "stone fleet" sunk in the main channel, which at that time had twelve feet of water at low tide where the figure 7 indicates the present depth. There was, moreover, another channel, making out more to the southward, with nine feet of water, where the figure 3 indicates the present depth. The vessels were placed checker-wise, in such a manner as to impede navigation, while interfering least with the discharge of the water. The effect, nevertheless, was the formation of a shoal in a short time, and the scouring out of two channels, one on each side of the obstructions, through which twelve and fourteen feet can now be carried at low water. The increased water-way thus given to the ebb-tide caused it to abandon the old nine-foot channel on the less direct course to deep water. We have here the total obstruction of a channel which was of considerable importance to the southward trade by new conditions introduced at a point four miles distant from where the effect was produced, and we are warned how carefully all the conditions of the hydraulic system of a harbor must be investigated before undertaking to make any change in its natural conditions, lest totally unlooked-for results be produced at points not taken into consideration.

NEW YORK HARBOR.

Approaching now more closely to the consideration of the tidal conditions in New York Harbor, we will examine the progress of the tide-wave through Long Island Sound from the eastward to its meeting with that entering New York Bay at Sandy Hook.

TIMES AND HEIGHTS OF TIDES IN LONG ISD. SOUND AND NEW YORK HARBOR.

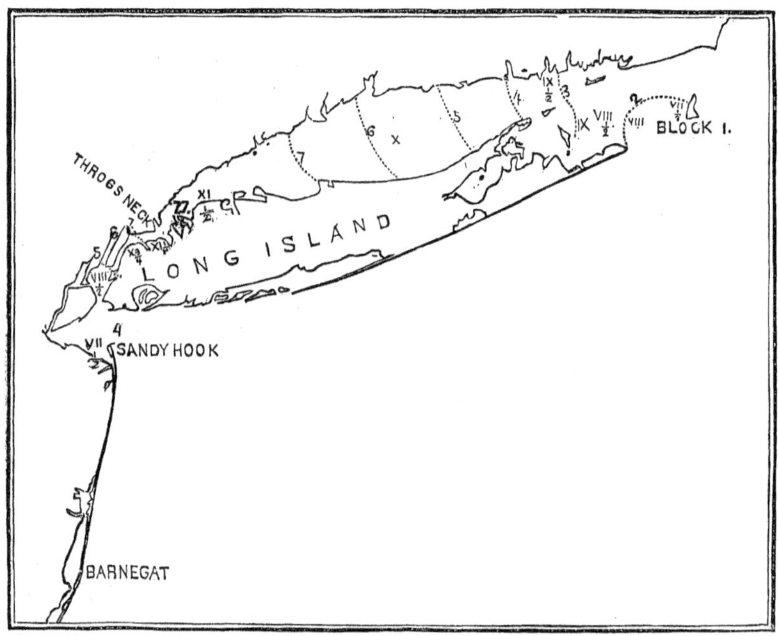

We see from the annexed diagram that about seven and a half hours after the transit of the moon high water has advanced just within Block Island with an elevation of two feet, and at the same time has just passed Sandy Hook with an elevation of four and a half feet. Traversing the sound at a rate indicated by the Roman figures, with increasing heights indicated by the Arabic numerals, it reaches Sand's Point eleven and a half hours after the transit of the moon with a height of seven and seven-tenths feet. The observed time of transmission from the Race to Sand's Point is two hours one minute, and the time computed from the depths, according to the law developed by Airy, is two hours fourteen minutes—a very good approximation, when we consider the irregularities in the configuration of the sound, which could not be taken into account. Advancing still farther, the height somewhat declines in consequence of the changes of direction in the channel and its shallowness. At Hell Gate this tide-wave is met by that which had entered at Sandy Hook, and advanced more slowly, owing to the narrowness and intricacies of the channel, especially in the East River.

These two tides, which meet and overlap each other at Hell Gate, differing from each other in times and heights, cause contrasts of water-elevations between the sound and harbor which call into existence the violent currents that traverse the East River. The conditions of the tidal circulation through Hell Gate are such, that if there were a partition across it, the water would sometimes stand nearly five feet higher, and at other times five feet lower, on one side than on the other. In the actual case of the superposition or compounding of the two tides the difference of level existing at any time is of course much less; but the difference of one foot is often observed within the space of 100 feet in the most contracted portion of Hell Gate, off Hallett's Point Referring now more particularly to the diagram representing New York Bay and Harbor, it is important to observe that the entrance from Long Island Sound is a natural depression or arm of the sea, which is not changed by the forces now in operation. The tidal currents which flow through it do not change the channel, but are obliged to follow it in its tortuous course. The Sandy Hook entrance, on the contrary, is characterized by a cordon of sands, extending from Sandy Hook to Coney Island, intersected by channels, which are maintained against the action of the sea, that tends to fill them up, by the scour of the ebb-tide from the tidal basin of New York Harbor.

Unlike Hell Gate passage, where permanence is the leading characteristic, the bar and channels of Sandy Hook have undergone continual changes within the brief period of our history. The advance of Sandy Hook upon the main ship-channel is among the notable and important instances of the effect of tidal currents. Within a century it has increased a mile and a quarter. In the place where the beacon on the end of the Hook now stands were forty feet of water fifteen years before it was built. The cause of this growth is a remarkably northwardly current along both shores of the Hook, running both during the flood and the ebb tides with varying rates, and resulting from those tides directly and indirectly.

The best water over the bar is about two miles east of Sandy Hook light, in a direct line with the swash channel, which is the second opening—shown on the sketch—above the Hook; the shoal lying between the main or Hook channel and the swash channel being known as Flynn's Knoll. The greatest depth over the bar is twenty-two feet at mean low water; and very nearly the same depth can now be carried through the swash channel, which formerly was three feet shallower,

ENTRANCE TO NEW YORK HARBOR

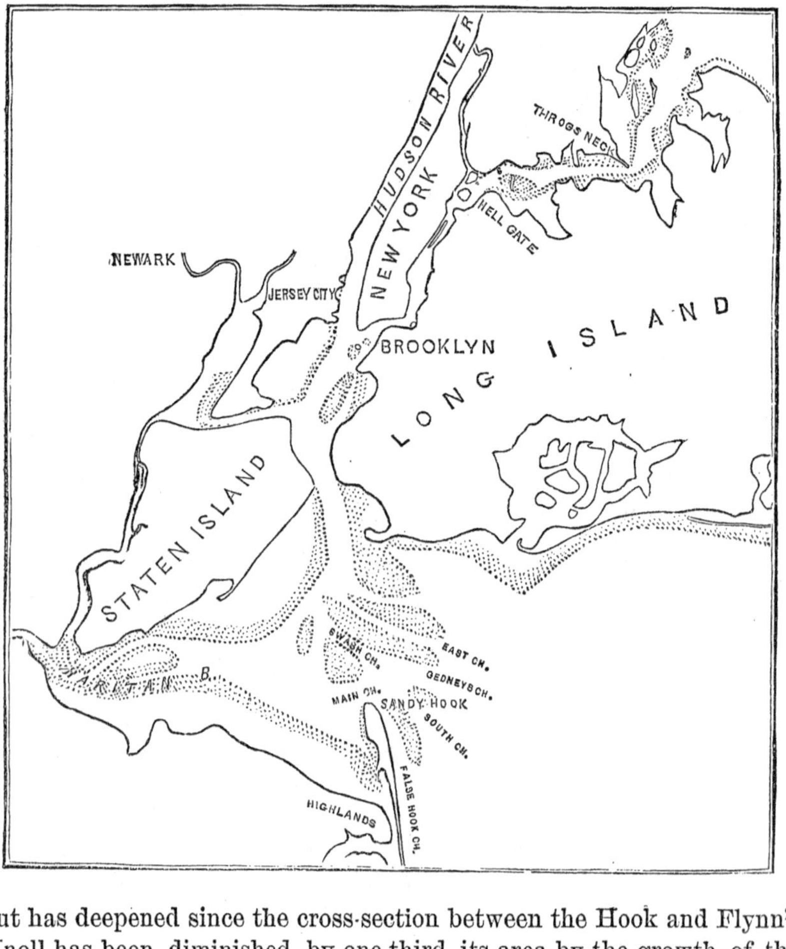

but has deepened since the cross-section between the Hook and Flynn's Knoll has been diminished by one-third its area by the growth of the Hook. This relative change in the capacity of the channels has not, however, affected the depth on the outer bar, which, according to the principles above laid down, is dependent mainly upon the area of the tidal basin within.

The depth of twenty-two feet at mean low water, which is now maintained at the entrance through the sands constantly thrown up by the waves of the sea, may be considered as depending upon the following elements:

1st. The large basin between Sandy Hook and Staten Island, including Raritan Bay, which furnishes more than one-half of the whole ebb-scour;

2d. What is called the Upper Bay, including the Jersey flats and Newark Bay;

3d. The North River, perhaps as far as Dobbs' Ferry, maintaining the head of the ebb-current, although not directly taking part in the outflow; and,

4th. A portion of the sound tide, which flows in through Hell Gate.

The proportion of the three first divisions in producing the depth of channel may be approximately estimated by a comparison of the areas and distances from the bar. In order to maintain the depth which we now have, it is important that the area of the tidal basin should not be encroached upon. In proportion as that is diminished the depth of the channels will decrease.

The flats, just bare at low water, but covered at high tide, form as important a part as any other portion, for it is obvious that it is only the volume of water contained between the planes of low and high water—the "tide-prism"—that does the work in scouring the channels. The water on the flats is especially useful by retarding the outflow, thus allowing a greater difference of level to be reached between the basin and the ocean.

When we yield to the demands of commerce any portion of the tidal territory to be used for its wharves and docks, we must do so with full cognizance of the sacrifice we are about to make in the depth of water over the bar; and in order to form any well-founded judgment in regard to the effect of such encroachments, it is necessary to be in possession of the fullest knowledge of all the physical facts involved in the problem, and no measure of encroachment should be determined upon except in pursuance of the advice of scientific experts.

A proposition frequently mooted by men of enterprise, and resisted by those interested in the welfare of the city of New York, is the occupation of the Jersey flats from Paulus Hook to Robbins Reef for docks and wharves. Without expressing any opinion as to the relative value of the gain of accommodation for shipping and the loss of depth in the channel, I venture to say that the withdrawal of that area from the domain of the tide would occasion a loss of not less than one foot in the depth of the bar off Sandy Hook, and certainly not more than two feet.

The part which the fourth division in our classification of the basin of New York, that of the East River and Hell Gate passage, plays in the outflow of the ebb-tide through the Sandy Hook channels, depends less upon the area involved than upon the difference in point of time and height of tide in Hell Gate, already adverted to. The westerly current, usually called the ebb-stream, since it falls in with the ebb-stream of New York Harbor, taking place when the sound-tide is highest, starts from a level of three and a half feet higher than the easterly, and thus a much larger amount of water flows out through the Sandy Hook channels than through the narrows at Throg's Neck. It is apparent, then, that this portion of the ebb-stream, re-enforcing as it does the ebb-stream of the harbor proper at the most favorable times, performs a most important part in maintaining the channels through the Sandy Hook bar. It may be estimated that the closing of Hell Gate would cause the loss of certainly not less than three feet in the depth of those channels.

From what has been said with regard to the meeting of the tides in Hell Gate, it will be seen that the violent currents experienced in that locality are due to causes beyond our control. The dangers to navigation arising from these currents, however, by their setting vessels upon the rocks and reefs, may, in a great measure, be done away with by the removal of the obstructions, in which work considerable progress has already been made. The removal of the reef at Hallett's Point, the work upon which is now in progress, will doubtless, in a great degree, do away with the eddies and under-currents produced by the sharp turn which the channel now takes at that point. It is not improbable that the successful removal of those obstructions will yet cause the sound entrance to be used in preference to the other by the fleets plying between European ports and the great commercial metropolis of America.

NOTE.—The reader who wishes to enter upon the mathematical treatment of the subject of tides is referred to Airy's treatise on tides and waves, and to the memoirs of Whewell and Lubbock, in the Philosophical Transactions of the Royal Society; and for investigations of the laws of the tides on our own coasts, to the papers on that subject by Bache and others in the annual reports of the Coast Survey. Among the latter, the lecturer is particularly indebted to the "Report on the tides and currents of Hell Gate," by Henry Mitchell, 1867, in which the complicated problem of the tidal circulation of New York Harbor is treated with great ability and success.

www.ingramcontent.com/pod-product-compliance
Lightning Source LLC
LaVergne TN
LVHW011146110826
845150LV00008B/2544
* 9 7 8 1 4 1 8 1 9 2 2 4 2 *